Journey to Europa

The NASA *Europa Clipper* Mission

James M. Thomas

Journey to Europa: The NASA *Europa Clipper* Mission.
Copyright © 2024 James M. Thomas
All rights reserved.
ISBN: 9798334612846

Cover Art: Artist rendering of NASA *Europa Clipper* over Europa with Jupiter in the distance. Image Credit: NASA/JPL.

Contents

Introduction 1
The Purpose of the *Europa Clipper* Mission 5
The *Europa Clipper* Spacecraft 11
The *Europa Clipper* Mission Timeline 21
Experiments, and Scientific Goals 25
    EIS 25
    E-THEMIS 28
    Europa-UVS 30
    MISE 32
    ECM 34
    PIMS 36
    Gravity/Radio Science 38
    REASON 40
    MASPEX 42
    SUDA 44
History of Jupiter System Exploration 49
The Future of Europa Exploration 53
Conclusion 59
Appendices 63
    Detailed Mission Timeline 65
    References and Further Reading 71
References 75
About the Author 77

Introduction

The quest to explore the outer reaches of our solar system has always captivated human imagination, driven by an insatiable curiosity to understand our place in the universe. Among the myriad celestial bodies that populate our solar system, Jupiter and its moons have stood out as objects of profound interest, holding secrets that could reshape our understanding of planetary science and the potential for life beyond Earth. The NASA *Europa Clipper* mission, set to embark on a historic journey to one of Jupiter's most intriguing moons, Europa, represents a bold step forward in this grand scientific endeavor.

The Significance of Europa

Europa has long fascinated scientists and astronomers. First glimpsed through Galileo's telescope in 1610, Europa is slightly smaller than our Moon and covered with a layer of ice. However, it is what lies beneath this icy shell that has piqued the curiosity of scientists worldwide. Evidence suggests that beneath its frozen crust, Europa harbors a vast ocean of liquid water, kept warm by tidal forces generated by Jupiter's immense gravitational pull. This subsurface ocean, potentially twice the volume of all Earth's oceans combined, makes Europa one of the most promising places in our solar system to search for extraterrestrial life.

The Quest for Extraterrestrial Life

The search for life beyond Earth is one of the driving forces behind modern planetary exploration. Life as we know it requires three basic ingredients: liquid water, an energy source, and a suite of chemical building blocks. Europa appears to have all three. The ocean beneath its icy shell is in contact with the moon's rocky mantle, providing the necessary chemical interactions that could sustain life. The discovery of potential plumes of water vapor erupting from Europa's surface has further fueled excitement, suggesting that sampling the subsurface ocean might be possible without the need to drill through miles of ice.

Mission Goals and Objectives

The primary goal of the *Europa Clipper* mission is to determine whether there are habitable conditions within Europa's subsurface ocean. The spacecraft will conduct a detailed survey of Europa's ice shell and subsurface ocean, analyze the composition of the moon's surface, and investigate its geology and potential for recent activity. By flying close to Europa approximately 50 times over its mission duration, *Europa Clipper* will gather high-resolution images and data, providing unprecedented insights into this enigmatic world.

Building on Previous Missions

The *Europa Clipper* mission stands on the shoulders of giants. It builds upon the legacy of previous missions that have explored the Jupiter system, including the *Pioneer* and *Voyager* missions, which provided the first close-up images of Jupiter and its moons. The *Galileo* orbiter, which conducted detailed studies of the Jovian system from 1995 to 2003, revealed compelling evidence of Europa's subsurface ocean and active geology. More recently, the *Juno* mission has been studying Jupiter's atmosphere and magnetosphere, offering valuable data that will aid *Europa Clipper*'s mission planning.

A New Era of Exploration

With the October 14, 2024-launch of *Europa Clipper*, we stand at the threshold of a new era in the exploration of our solar system. This mission not only aims to answer fundamental questions about Europa's potential to support life but also serves as a precursor to future missions that may one day directly explore its surface or even delve into its ocean depths. The discoveries made by *Europa Clipper* will not only expand our knowledge of Europa but also inspire future generations to look to the stars and continue humanity's quest to explore the unknown.

The Purpose of the *Europa Clipper* Mission

Scientific Objectives of the Mission

The primary objective of the NASA *Europa Clipper* mission is to assess the habitability of Jupiter's moon, Europa. This icy moon has captured the scientific community's attention due to compelling evidence suggesting that it harbors a subsurface ocean beneath its frozen crust. The mission aims to determine whether the conditions within this hidden ocean could support life.

1. **Search for Habitable Conditions:**
   Europa Clipper is designed to investigate the essential elements that could make Europa habitable. The mission will focus on three key factors:

   - **Liquid Water**: Confirming the presence and properties of the subsurface ocean.

   - **Chemistry**: Identifying essential chemical elements and compounds.

   - **Energy**: Understanding the sources of energy that could support biological processes.

2. **Study the Ice Shell and Subsurface Ocean:**
   The spacecraft will use a suite of sophisticated instruments to measure the thickness of Europa's ice shell and the depth of its ocean. By examining the ice-ocean interface, scientists hope to understand how the ocean interacts with the moon's surface and whether it could support life.

3. **Analyze Surface Composition:**
   Europa Clipper will analyze the composition of Europa's surface to identify organic compounds and other key chemical constituents. This analysis will help determine the

moon's potential for supporting life and provide insights into its geochemical processes.

4. **Investigate Geological Activity:**
   The mission will study Europa's geology to understand its history and current state. This includes examining surface features such as ridges, cracks, and chaos terrain, which are believed to be influenced by the subsurface ocean.

Importance of Europa in the Search for Extraterrestrial Life

Europa stands out in the search for life beyond Earth due to its subsurface ocean, which may be one of the largest in the solar system. The presence of liquid water is crucial because all known life forms on Earth require it. Europa's ocean is thought to be in contact with its rocky mantle, potentially allowing chemical interactions similar to those on Earth that support life.

Furthermore, evidence of water vapor plumes erupting from Europa's surface suggests that material from the ocean could be sampled without drilling through the ice, making it an accessible target for study. These plumes provide an opportunity to analyze the ocean's composition and search for signs of life directly.

Mission Goals and Objectives

The *Europa Clipper* mission is structured around several core scientific goals:

1. **Characterize the Ice Shell and Ocean:**

   - Determine the ice shell's thickness and its variability.

   - Measure the depth and salinity of the subsurface ocean.

- Investigate the ocean's dynamics and its potential for harboring life.

2. **Investigate Surface and Subsurface Composition:**

- Map the distribution of organic and inorganic compounds on the surface.
- Identify areas with recent or ongoing geological activity.
- Analyze the composition of plumes, if present.

3. **Study Geological Features and Processes:**

- Understand the formation and evolution of surface features.
- Examine the role of tidal forces in shaping Europa's geology.
- Assess the potential for current geological activity.

By achieving these goals, *Europa Clipper* aims to provide a comprehensive understanding of Europa's potential habitability and pave the way for future exploration missions, including potential landers and probes that could directly sample the moon's surface or ocean.

Importance in the Broader Context of Space Exploration

The *Europa Clipper* mission is a crucial step in humanity's quest to explore the outer solar system and search for life beyond Earth. It builds on the legacy of previous missions to Jupiter and its moons, incorporating advanced technology and scientific knowledge.

1. **Building on Previous Discoveries:**

    - *Galileo* **Mission**: Provided strong evidence for Europa's subsurface ocean and revealed its complex geology.

    - *Voyager* **Missions**: Offered the first detailed images of Europa's icy surface.

    - *Juno* **Mission**: Currently studying Jupiter's atmosphere and magnetic field, providing data that will complement *Europa Clipper*'s findings.

2. **Technological Advancements:**
Europa Clipper will employ cutting-edge technology, including high-resolution cameras, ice-penetrating radar, and sophisticated spectrometers, to conduct its investigations. These instruments represent significant advancements over those used in previous missions.

3. **Future Exploration:**
The mission's findings will inform future exploration efforts, including potential landers and subsurface probes. Understanding Europa's habitability could guide the design and objectives of these missions, ultimately bringing us closer to answering the question of whether life exists elsewhere in the solar system.

A New Era in Planetary Science

As we embark on the *Europa Clipper* mission, we are entering a new era in planetary science and exploration. The mission's success could redefine our understanding of habitability and the potential for life beyond Earth. The discoveries made by *Europa Clipper* will not only expand our knowledge of this enigmatic moon but also inspire future generations to continue the quest for knowledge and exploration.

Mission patch of the NASA Europa Clipper *mission. Image Credit: NASA/JPL.*

The *Europa Clipper* mission is managed by Caltech in Pasadena, California. NASA's Jet Propulsion Laboratory (JPL) leads the development of the *Europa Clipper* mission in partnership with the Johns Hopkins Applied Physics Laboratory (APL) in Laurel, Maryland, for NASA's Science Mission Directorate in Washington. The main spacecraft body was designed by APL in collaboration with JPL and NASA's Goddard Space Flight Center in Greenbelt, Maryland. The Planetary Missions Program Office at NASA's Marshall Space Flight Center in Huntsville, Alabama, executes program management of the *Europa Clipper* mission. NASA's Launch Services Program, based at Kennedy, manages the launch service for the *Europa Clipper* spacecraft.

In the following chapters, we will delve deeper into the details of the *Europa Clipper* mission, exploring the spacecraft's design, the scientific instruments it carries, the timeline of its journey, and its broader significance in the context of space exploration. Join us as we embark on this exciting journey to one of the most promising worlds in our solar system.

The *Europa Clipper* Spacecraft

The *Europa Clipper* spacecraft represents a marvel of modern engineering and scientific innovation, designed to explore one of the most intriguing bodies in our solar system. In this chapter, we delve into the spacecraft's design, its key components, and the sophisticated instruments it carries to achieve its ambitious scientific goals.

Artist's rendering of the Europa Clipper *spacecraft. Image Credit: NASA/JPL.*

Design and Engineering

The *Europa Clipper* spacecraft is one of the largest and most complex interplanetary spacecraft ever built by NASA, and the largest spacecraft ever built for planetary exploration. Its design reflects the mission's unique requirements, balancing the need for robustness in the harsh environment of space with the precision required for scientific investigation.

Spacecraft Structure

Dimensions: Approximately 16 feet (5 meters) tall, with a wingspan (solar array tip-to-tip) of about 100 feet (30.5 meters), longer than a basketball court.

Mass: The total mass of the spacecraft at launch was around 6,000 kilograms (13,227 pounds), including propellant.

Configuration: The spacecraft is a modular structure designed for durability and efficiency, featuring a central body with attached solar arrays, a high-gain antenna, and various scientific instruments.

Bus: The central structure, or bus, of the spacecraft houses essential systems such as power, propulsion, and thermal control. It is constructed from lightweight, high-strength materials to withstand the rigors of space travel.

Radiation Shielding: Given Jupiter's intense radiation environment, *Europa Clipper* includes substantial radiation shielding to protect its electronics and instruments. This shielding is vital to ensure the spacecraft's longevity and functionality.

Power System

Solar Arrays: The spacecraft is powered by two large solar arrays, spanning over 100 feet when fully deployed, and generating up to 600 watts of power. These arrays are essential for operating the spacecraft's systems and scientific instruments during its mission.

Batteries: High-capacity batteries store energy generated by the solar arrays, ensuring continuous power supply during periods of darkness or high-power demand.

Propulsion System

Main Engine: The spacecraft is equipped with a bi-propellant main engine for major trajectory adjustments and orbital insertion around Jupiter.

Reaction Control System: A series of smaller thrusters provide fine-tuned control for attitude adjustments and small course corrections.

Navigation: *Europa Clipper* will rely on precise navigation to perform close flybys of Europa. It uses star trackers, gyroscopes, and a sophisticated guidance system to maintain its course and orientation.

Communication System

High-Gain Antenna: The spacecraft is equipped with a high-gain antenna to transmit data back to Earth. This antenna enables high-bandwidth communication, essential for sending large volumes of scientific data.

Low-Gain Antennas: These are used for low-rate communication, primarily during initial phases of the mission and for redundancy.

Navigation System

Function: Utilizes a combination of star trackers, gyroscopes, and accelerometers to maintain precise navigation and orientation.

Capabilities: Ensures accurate targeting for flybys and scientific observations, as well as efficient use of propulsion for trajectory adjustments.

Thermal Control System

Thermal Blankets and Radiators: The spacecraft is equipped with thermal blankets and radiators to manage its temperature. These systems ensure that the spacecraft's instruments and electronics operate within their optimal temperature ranges.

Radiation Protection

Radiation Shielding: Aluminum and Titanium shielding protect sensitive electronics and instruments from Jupiter's intense radiation environment. The shielding also ensures long-term functionality of the spacecraft and instruments by minimizing radiation-induced damage.

Radiation-Hardened Electronics: The spacecraft utilizes specially designed components that can withstand high levels of radiation. The components enhance the durability and reliability of the spacecraft's systems during its mission around Jupiter.

Scientific Instruments on Board

The *Europa Clipper* carries an impressive array of scientific instruments, each designed to probe different aspects of Europa's environment. These instruments will provide a comprehensive understanding of the moon's ice shell, subsurface ocean, surface composition, and potential habitability.

Wide-Angle and Narrow-Angle Cameras (EIS):

   Objective: To capture high-resolution images of Europa's surface.

   Function: These cameras will provide detailed maps of the surface, revealing geological features and potential landing sites for future missions.

Thermal Emission Imaging System (E-THEMIS):

Objective: To map the surface temperature of Europa and identify areas of recent geological activity.

Function: E-THEMIS uses infrared sensors to detect heat emitted from the surface, highlighting warmer regions that may indicate active processes.

Ultraviolet Spectrograph (Europa-UVS):

Objective: To study Europa's atmosphere and surface composition.

Function: Europa-UVS detects ultraviolet light, revealing details about the surface materials and any active plumes.

Mapping Imaging Spectrometer for Europa (MISE):

Objective: Map the distribution of ices, salts, organics, and the warmest hotspots of Europa.

Function: Analyzes infrared light reflected from Europa, measures the presence, absence, strength, and weakness of various wavelengths, or frequencies, of light.

Europa Clipper Magnetometer (ECM):

Objective: Provide data on the depth and salinity of the ocean, contributing to the understanding of its habitability.

Function: Measures the magnetic field around Europa to infer the properties of its subsurface ocean and its interaction with Jupiter's magnetosphere.

Plasma Instrument for Magnetic Sounding (PIMS):

Objective: To measure the interaction between Europa's atmosphere and Jupiter's magnetosphere.

Function: PIMS will analyze plasma waves and magnetic fields, providing data on the space environment around Europa.

Gravity/Radio Science:

Objective: Provide data on the moon's mass distribution, which will help to model the thickness of the ice shell and the properties of the subsurface ocean.

Function: Uses the spacecraft's communication system to measure Europa's gravitational field and infer details about its internal structure.

Radar for Europa Assessment and Sounding: Ocean to Near-surface (REASON):

Objective: Detect features within the ice shell up to several kilometers deep, helping to determine the thickness of the ice and the properties of the underlying ocean.

Function: Uses radar waves to penetrate Europa's icy crust and map its subsurface structure, including the suspected ocean beneath the ice.

MAss Spectrometer Planetary Exploration/Europa (MASPEX):

Objective: Identify organic molecules and other compounds, contributing to the search for biosignatures.

Function: Analyzes the composition of Europa's thin atmosphere and any material ejected from its surface, such as plumes.

Surface Dust Analyzer (SUDA):

Objective: Analyze dust particles ejected from Europa's surface.

Function: Study dust particles and identify the composition of Europa's surface and any subsurface materials brought to the surface by geological activity.

How Each Instrument Contributes to Mission Objectives

Each of these instruments plays a crucial role in addressing the mission's scientific objectives. Together, they will provide a comprehensive dataset that helps answer fundamental questions about Europa's habitability:

Europa Imaging System (EIS): A wide-angle camera and a narrow-angle camera, each with an eight-megapixel sensor, will produce high-resolution color and stereoscopic images of Europa. They will study geologic activity, measure surface elevations, and provide context for other instruments.

Europa Thermal Emission Imaging System (E-THEMIS): The thermal imager uses infrared light to distinguish warmer regions on Europa where warm liquid water may be near the surface or might have erupted onto the surface. It will also measure surface texture to understand the small-scale properties of the surface.

Europa Ultraviolet Spectrograph (Europa-UVS): By collecting ultraviolet light with a telescope, and creating images, the mission's ultraviolet spectrograph will help determine the composition of Europa's atmospheric gases and surface materials. It will also search near Europa for signs of plume activity.

Mapping Imaging Spectrometer for Europa (MISE): The mission's infrared spectrometer will map the distribution of ices, salts, organics, and the warmest hotspots on Europa. The maps will help scientists understand the moon's geologic history and determine if Europa's suspected ocean is suitable for life.

Europa Clipper Magnetometer (ECM): The magnetometer investigation aims to confirm that Europa's ocean exists, measure its depth and salinity, and measure the moon's ice shell thickness. It will also study Europa's ionized atmosphere and how it interacts with Jupiter's ionized atmosphere.

Plasma Instrument for Magnetic Sounding (PIMS): Europa's ionosphere, and plasma trapped in Jupiter's magnetic field, distort magnetic fields near Europa. PIMS Faraday cups will distinguish those distortions from Europa's induced magnetic field, which carries information about Europa's ocean.

Gravity/Radio Science: Europa and its gravity field flex as the moon's non-circular orbit carries it closer, then farther, from Jupiter. Measuring Europa's gravity at various points in the moon's orbit will show how Europa flexes and help reveal its internal structure.

Radar for Europa Assessment and Sounding: Ocean to Near-surface (REASON): Ice-penetrating radar will probe Europa's icy shell for the moon's suspected ocean and study the ice's structure and thickness. It will also study the moon's surface elevations, composition, and roughness, and search the moon's atmosphere for plumes.

MAss Spectrometer for Planetary EXploration/Europa (MASPEX): The mass spectrometer will analyze gases in Europa's faint atmosphere and possible plumes. It will study the chemistry of the moon's suspected subsurface ocean, how ocean and surface exchange material, and how radiation alters compounds on the moon's surface.

SUrface Dust Analyzer (SUDA): Tiny meteorites eject bits of Europa's surface into space, and a subsurface ocean or reservoirs might vent material into space as plumes. The dust analyzer will identify that material's chemistry and area of origin, and offer clues to Europa's ocean salinity.

By combining data from these instruments, scientists hope to gain a holistic understanding of Europa, its potential to support life, and the processes that shape its enigmatic environment.

The *Europa Clipper* Mission Timeline

The *Europa Clipper* mission has been years in the making, marked by significant milestones from its initial conception to its forthcoming journey to Jupiter's moon, Europa. This chapter provides a detailed timeline of the mission's development, highlighting the key events and stages that have brought us to the brink of this historic exploration.

Development and Planning Stages

Initial Proposals and Mission Design (2008-2013):

- The idea for a dedicated mission to Europa emerged from the scientific community's growing interest in exploring icy moons with subsurface oceans. Initial studies and proposals were developed, focusing on the feasibility of such a mission.

- By 2011, NASA began formal studies and mission concept reviews, which led to the selection of the *Europa Clipper* concept in 2013. This concept was chosen due to its innovative approach to conducting multiple flybys of Europa, minimizing exposure to Jupiter's intense radiation.

Selection and Funding (2014-2015):

- In 2014, NASA officially selected the *Europa Clipper* mission for further development. The mission received initial funding for preliminary design and technology development.

- In 2015, the mission passed its first major review, known as Key Decision Point-B (KDP-B), allowing it to proceed into the formulation phase. Congress provided additional funding, emphasizing the mission's importance.

Key Milestones Leading Up to Launch

Construction and Testing Phases (2016-2023):

- **2016-2018**: The mission entered the preliminary design phase, during which engineers and scientists developed detailed plans for the spacecraft and its instruments. Key technologies were tested and validated to ensure they could withstand the harsh conditions of space and Europa's environment.

- **2019**: The mission passed its Critical Design Review (CDR), confirming that the design was sound and ready for construction.

- **2020-2021**: Assembly of the spacecraft began, with components being integrated and tested at NASA's Jet Propulsion Laboratory (JPL). This phase included extensive testing of individual systems and instruments.

- **2022-2023**: The completed spacecraft underwent environmental testing to simulate the conditions of launch, space travel, and operations around Jupiter. This included vibration tests, thermal vacuum tests, and electromagnetic interference tests.

Environmental and System Testing at JPL (2023-2024):

- In early 2023, the spacecraft entered a rigorous testing phase at JPL. These tests ensured that all systems would function correctly under the extreme conditions of space.

- By mid-2024, the spacecraft had completed all major tests, demonstrating its readiness for the mission. This marked a critical milestone, paving the way for the spacecraft's transportation to the launch site.

Transportation to Kennedy Space Center (May 2024):

- On May 23, 2024, the *Europa Clipper* spacecraft arrived at NASA's Kennedy Space Center in Florida. This event marked the beginning of the final preparations for launch, including integration with the launch vehicle and final system checks.

Launch and Journey to Jupiter

Launched October 14, 2024, at 1:26 p.m. EDT:

- Launched aboard a SpaceX Falcon Heavy rocket from Launch Pad 39A at Kennedy Space Center, the spacecraft begins its journey to Jupiter, marking a major milestone in NASA's exploration of the outer solar system. The successful launch is the culmination of years of preparation, and it sets the stage for one of the most important scientific missions of the coming decade.

Initial Spacecraft Checkout and Instrument Activation (2024-2025):

- Following the launch, *Europa Clipper* undergoes a series of post-launch tests and checks, ensuring all systems and scientific instruments are fully operational as it travels through interplanetary space. This phase ensures that the spacecraft is in full working order for its long journey to Jupiter.

Mars Flyby, March 1, 2025, at 12:57 p.m. EST:

- At that moment, the spacecraft was roughly 550 miles (884 km) above the Martian surface, moving at about 15.2 miles/sec (24.5 km/sec) relative to the Sun. This point marked the peak of the spacecraft's gravity-assist maneuver, which also doubled as a crucial moment to test thermal imaging (E-THEMIS) and the full radar system in flight.

Earth Fly-by, December 3, 2026, at around 4:15 p.m. EST

- This maneuver will bring *Europa Clipper* within approximately 3,200 km (about 2,000 miles) of Earth's surface—close enough to reshape its trajectory and provide a critical velocity boost for the rest of its journey. The Earth flyby serves as the **second and final gravity assist** in the mission's inner solar system leg, imparting the extra energy needed to reach Jupiter with minimal fuel usage.

Cruise to Jupiter (2025-2030):

- Following the Earth fly-by, the spacecraft will follow a high-energy arc outward, eventually arriving at Jupiter in April 2030, where it will enter orbit and commence detailed Europa flybys.

Mission Phases Upon Arrival

Orbit Insertion Around Jupiter (April 2030):

- Upon arrival, the spacecraft will perform a critical maneuver to enter orbit around Jupiter. This phase will involve precise calculations and timing to ensure the spacecraft is captured by Jupiter's gravity without being pulled too close to the planet.

Multiple Flybys of Europa (2030-2035):

- *Europa Clipper* will conduct approximately 50 flybys of Europa over several years. These flybys are strategically planned to maximize scientific return while minimizing radiation exposure. Each pass will bring the spacecraft close to Europa's surface, allowing its instruments to gather detailed data.

Data Collection and Transmission Back to Earth:

- Throughout the mission, the spacecraft will collect and transmit data back to Earth. This data will be analyzed by scientists to achieve the mission's scientific objectives. The spacecraft is equipped with high-gain antennas to facilitate high-bandwidth communication, ensuring that large volumes of data can be sent efficiently.

The *Europa Clipper* mission timeline encapsulates years of meticulous planning, development, and testing. Each milestone represents a step closer to unlocking the secrets of Europa and advancing our understanding of the potential for life beyond Earth. As we look forward to exploration, the mission's journey underscores the dedication and ingenuity required to embark on such a bold scientific endeavor.

Experiments, and Scientific Goals

The *Europa Clipper* mission is driven by a set of clearly defined scientific goals aimed at understanding the habitability of Jupiter's moon Europa. To achieve these goals, the spacecraft is equipped with a suite of sophisticated scientific instruments designed to conduct a variety of experiments. This chapter explores these instruments and the experiments they will perform, highlighting how each contributes to the mission's overarching objectives.

EIS
Wide-Angle and Narrow-Angle Cameras

Objective

Capture high-resolution images of Europa's surface.

Function

The Europa Imaging System (EIS, pronounced "ice") includes both wide-angle and narrow-angle cameras. The wide-angle camera (WAC) will provide context images of Europa's surface, while the narrow-angle camera (NAC) will capture detailed, high-resolution images of specific regions. The NAC pivots 60 degrees on two axes. Both cameras will produce stereoscopic images and have filters to acquire color images.

The NAC is a reflecting telescope and uses a large mirror to collect light. Then other mirrors and lenses direct a condensed beam onto an eight-megapixel detector called a complementary metal oxide semiconductor (commonly known as CMOS). Cell phones and digital cameras also use CMOS detectors, which are perhaps the most widely used NASA spinoff technology. The WAC has the same kind of sensor but is a refracting telescope instead of a reflecting telescope. It lets the light directly in and uses lenses to focus the light onto its detector.

EIS will map about 90% of Europa at 330 feet (100 meters) per pixel. That's six times more of Europa's surface than *Galileo*, at five times better resolution. When the spacecraft is close to Europa during flybys, EIS will produce images with a resolution 100 times better.

These images will be used to map surface features, study geological formations, and identify potential landing sites for future missions.

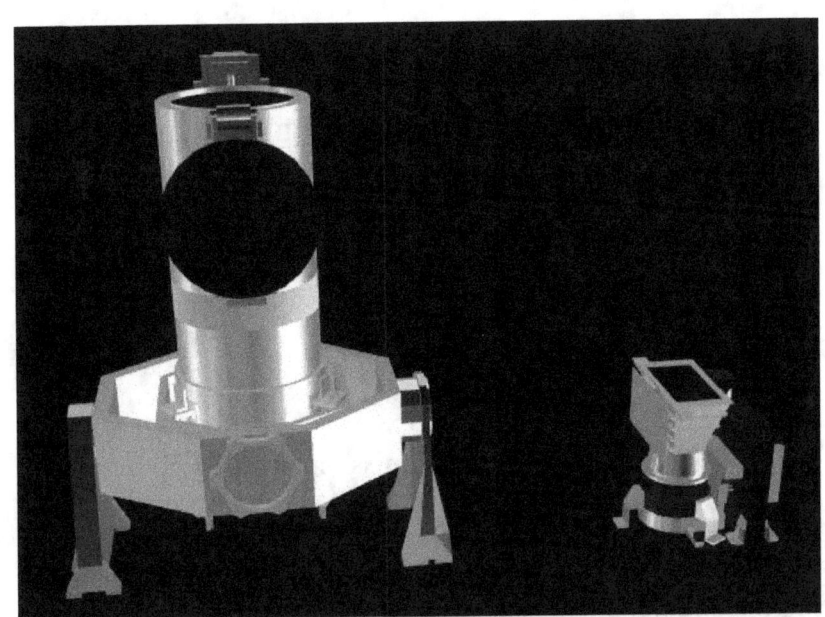

View 1 of Wide-Angle and Narrow-Angle Cameras (EIS). Image Credit: NASA.

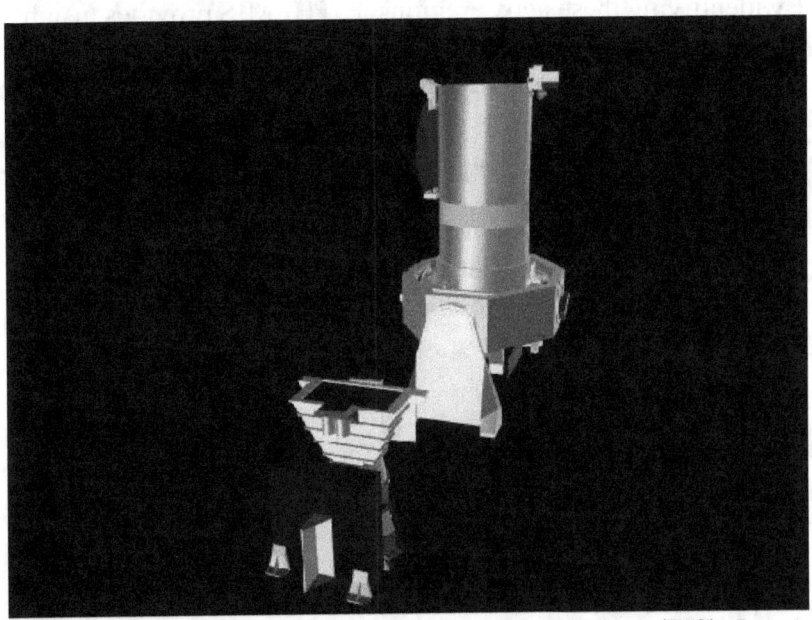

View 2 of Wide-Angle and Narrow-Angle Cameras (EIS). Image Credit: NASA.

E-THEMIS
Thermal Emission Imaging System

Objective

Map the surface temperature of Europa and identify areas of recent geological activity.

Function

E-THEMIS (Europa Thermal Emission Imaging System) will use infrared sensors to detect heat emitted from Europa's surface and map Europa's temperatures. Warmer areas may indicate regions of recent geological activity, such as cryovolcanos, or regions where the moon's suspected ocean may be near the surface.

By identifying these active regions, E-THEMIS provides insights into the geological processes shaping Europa's surface and helps prioritize areas for detailed study.

Like visible light, infrared light consists of photons. But infrared is outside humans' visual range. E-THEMIS directs infrared light through three filters. Each filter allows only certain wavelengths to pass through. The light then strikes a detector with three segments, each sensitive to one "color," or wavelength range, of light.

E-THEMIS will scan Europa's surface for relatively warm ice, which may be a sign of recent resurfacing. Other instruments can then target those areas to learn about the moon's subsurface chemistry. Warmer ice could also mean Europa's suspected ocean is closer to the surface in that location.

Also, when part of Europa rotates out of sunlight, thermal inertia causes granular material to cool faster than large blocks. E-THEMIS will record surface cooling rates to learn about the texture of Europa's surface. Mapping surface temperature and water near the surface will help to understand Europa's small-scale properties and might even help to find sites for a possible future lander.

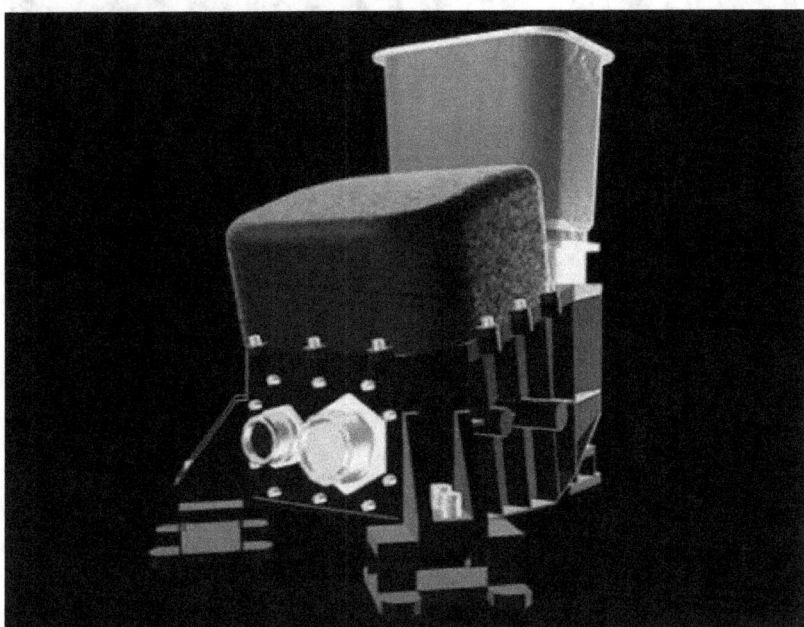

View 1 of Thermal Emission Imaging System (E-THEMIS). Image Credit: NASA.

View 2 of Thermal Emission Imaging System (E-THEMIS). Image Credit: NASA.

Europa-UVS
Ultraviolet Spectrograph

Objective

Study Europa's atmosphere and surface composition using ultraviolet light.

Function

Does Europa truly have a liquid water ocean beneath its icy shell? Does the ocean spew into space where the spacecraft can directly sample it? What materials other than ice are on Europa's icy surface? How does Jupiter's relentless radiation affect surface materials? Ultraviolet data will allow scientists to answer some of those questions by identifying Europa's materials more accurately than ever before.

Europa-UVS will detect ultraviolet light emitted and reflected by Europa's surface and atmosphere. It can identify the presence of water ice, salts, and organic compounds.

The instrument will also search for signs of plumes and analyze their composition, providing further evidence of subsurface ocean activity and potential habitability.

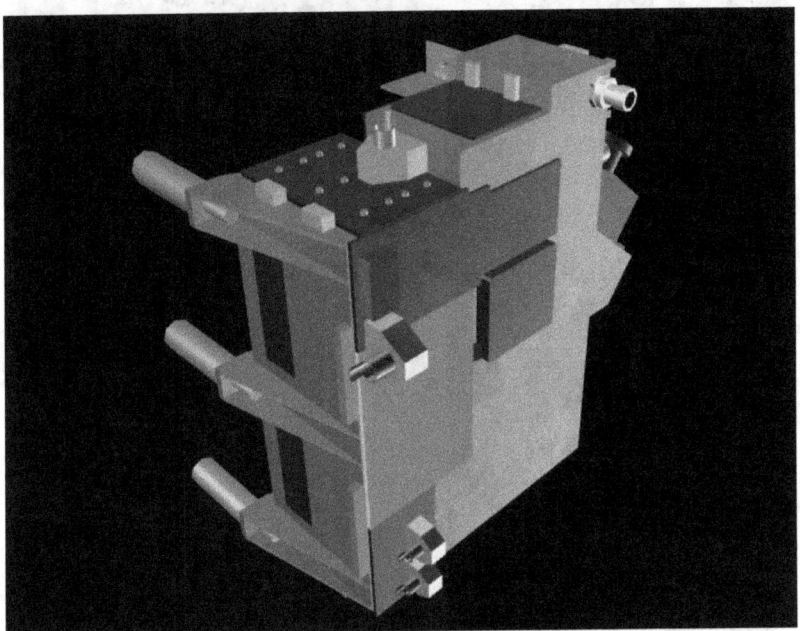

View 1 of Ultraviolet Spectrograph (Europa-UVS). Image Credit: NASA.

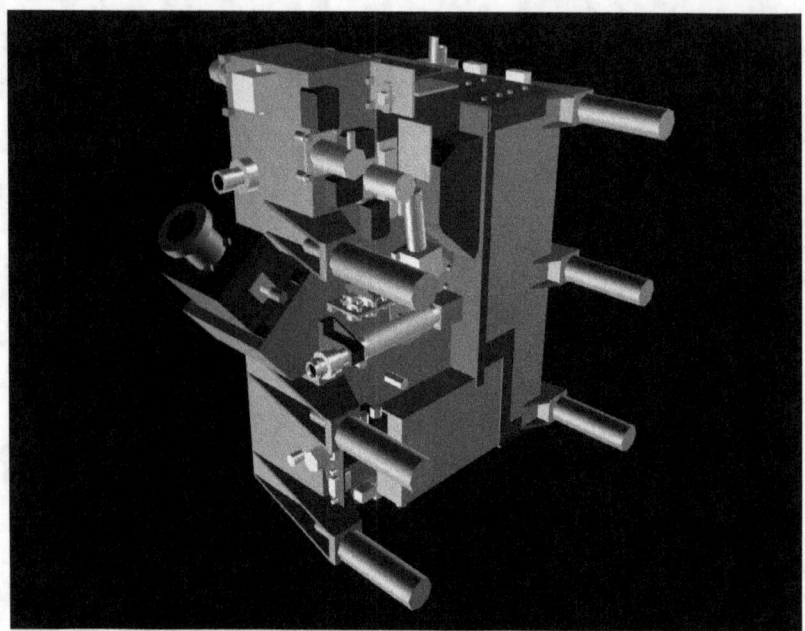

View 2 of Ultraviolet Spectrograph (Europa-UVS). Image Credit: NASA.

MISE
Mapping Imaging Spectrometer for Europa

Objective

Map the distribution of ices, salts, organics, and the warmest hotspots of Europa.

Function

The Mapping Imaging Spectrometer for Europa, or MISE (pronounced "mize"), will analyze infrared light reflected from Europa. It will measure the presence, absence, strength, and weakness of various wavelengths, or frequencies, of light. MISE will map Europa's surface composition in detail.

Infrared light enters MISE through a slit. Mirrors direct the light to a calcium fluoride (CaF_2) lens, which directs the light to a grating. The grating cuts light into discrete wavelengths, which then strike a detector. But MISE doesn't produce a picture all at once like a conventional camera.

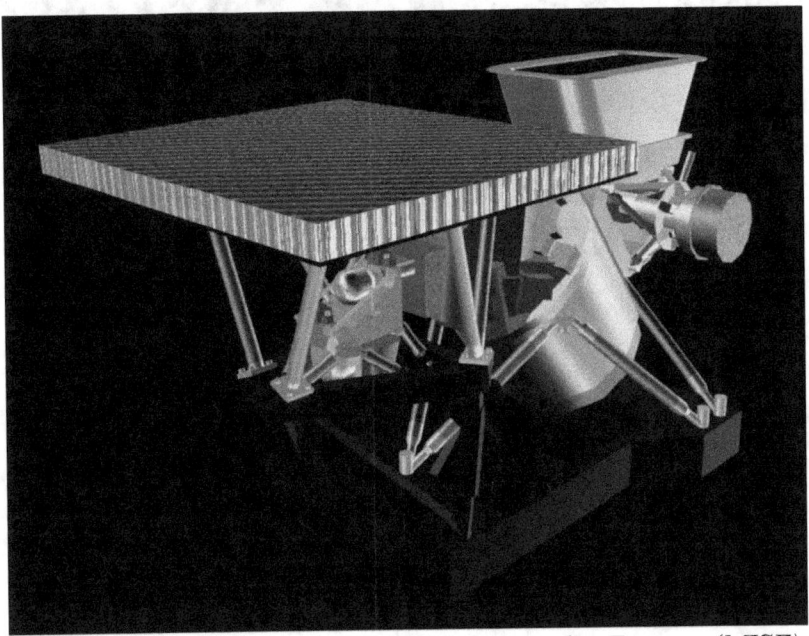

View 1 of Mapping Imaging Spectrometer for Europa (MISE). Image Credit: NASA.

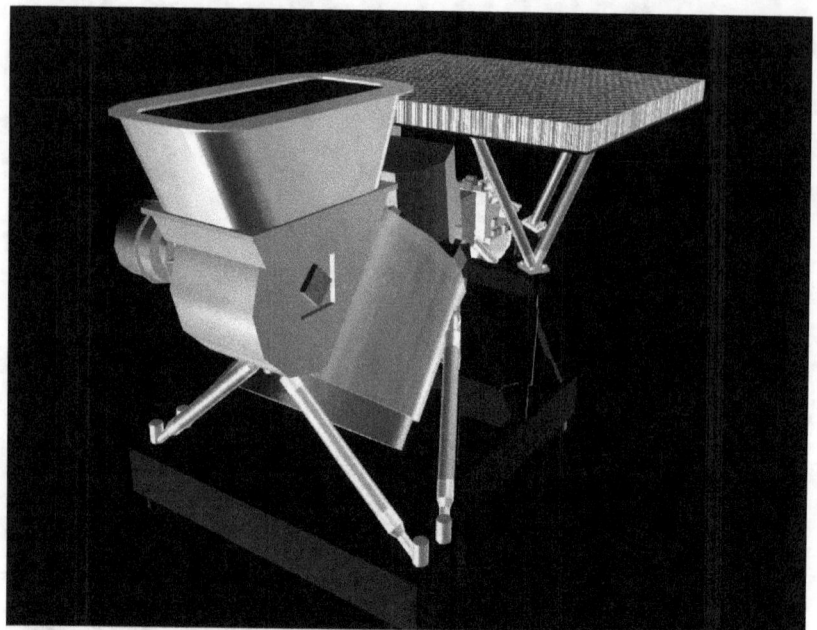

View 2 of Mapping Imaging Spectrometer for Europa (MISE). Image Credit: NASA.

ECM
Europa Clipper Magnetometer (ECM)

Objective

Study the magnetic field around Europa to infer the properties of its subsurface ocean.

Function

Jupiter rotates once every 10 hours, and its magnetic field spins right along with it. The magnetic field whips high-energy plasma along at almost the same rate. The plasma distorts the magnetic field near Europa, as does the spacecraft itself, and this affects magnetometer data.

But other science instruments will study plasma around Europa and learn how it affects the magnetic field. Engineers have modeled how the spacecraft will distort the magnetic field. The magnetometer team will remove those factors from their data to get a clearer and more accurate picture of Europa's magnetic field.

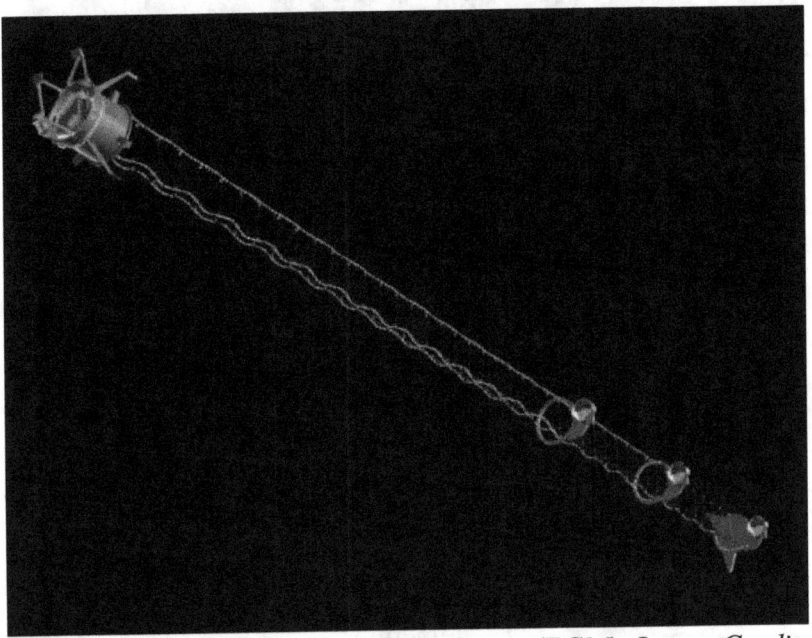

View 1 of Europa Clipper *Magnetometer (ECM). Image Credit: NASA.*

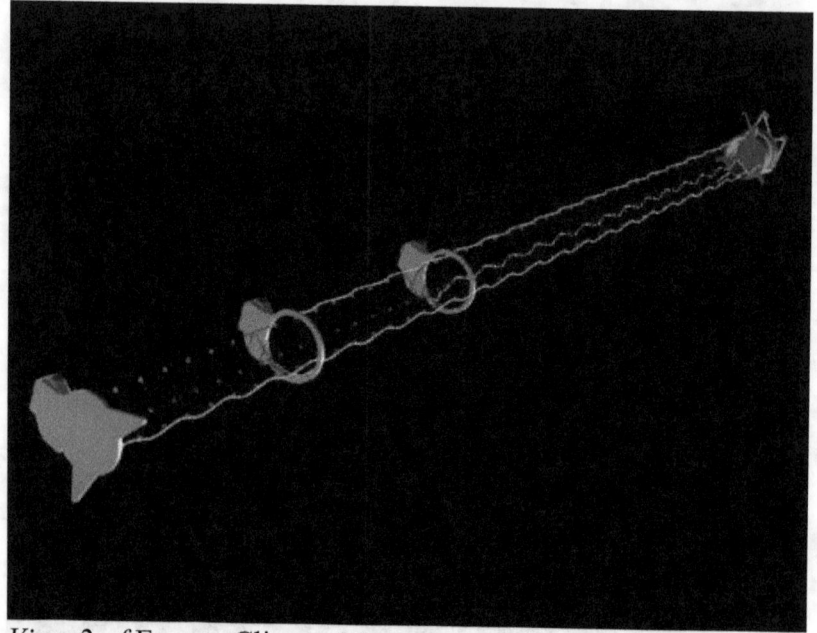

View 2 of Europa Clipper *Magnetometer (ECM). Image Credit: NASA.*

PIMS
Plasma Instrument for Magnetic Sounding

Objective

Measure the interaction between Europa's atmosphere and Jupiter's magnetosphere.

Function

PIMS (Plasma Instrument for Magnetic Sounding) will analyze plasma waves and magnetic fields in Europa's environment. This data helps understand how Europa's thin atmosphere interacts with the charged particles in Jupiter's magnetosphere.

By studying these interactions, PIMS provides insights into the space environment around Europa and the potential for atmospheric escape or surface modification by charged particles.

Plasma distorts magnetic fields around Europa and obscures the induction signal from the magnetometer. PIMS will allow scientists to model the plasma and subtract its contribution from magnetometer data, letting scientists study Europa's ocean depth and conductivity, and ice shell thickness.

View 1 of Plasma Instrument for Magnetic Sounding (PIMS). Image Credit: NASA.

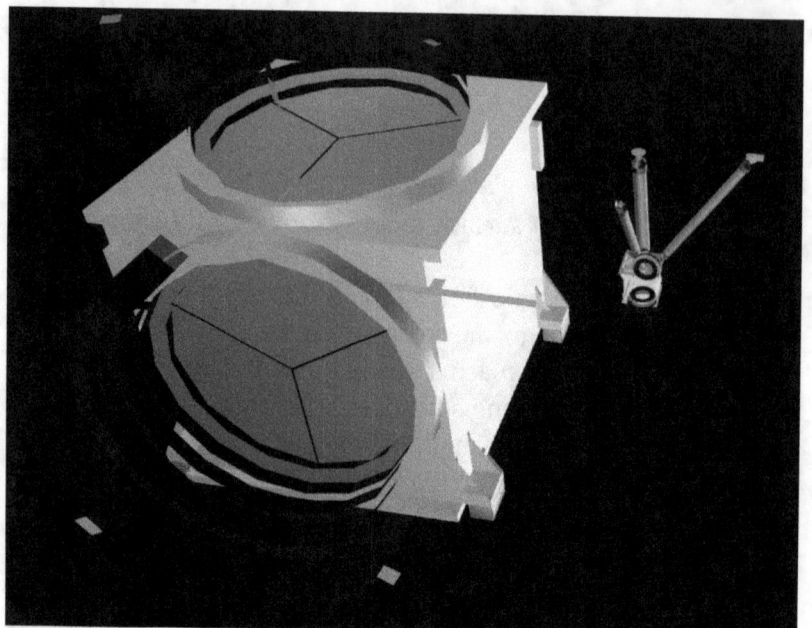

View 2 of Plasma Instrument for Magnetic Sounding (PIMS). Image Credit: NASA.

Gravity/Radio Science

Objective

Measure Europa's gravitational field to infer its internal structure.

Function

In gravity experiments, radio antennas on Earth send a radio signal to *Europa Clipper*. The spacecraft will then transmit to Earth at a frequency coherent to what it received. The gravity science team will precisely analyze the Doppler shift and other aspects of the radio signal received on Earth. Then they will trace it back a few steps.

Scientists will measure the difference between the frequency that the spacecraft sends, and the frequency Earth receives. The difference will give scientists details about the spacecraft's motion, which will give them details about Europa's gravity field. By studying how Europa's gravity field changes shape, scientists will be measuring how Europa changes shape. This will give them information about the moon's internal structure.

Understanding Europa's internal structure is crucial for assessing the thickness of the ice shell and the depth of the ocean, as well as identifying regions of potential subsurface activity.

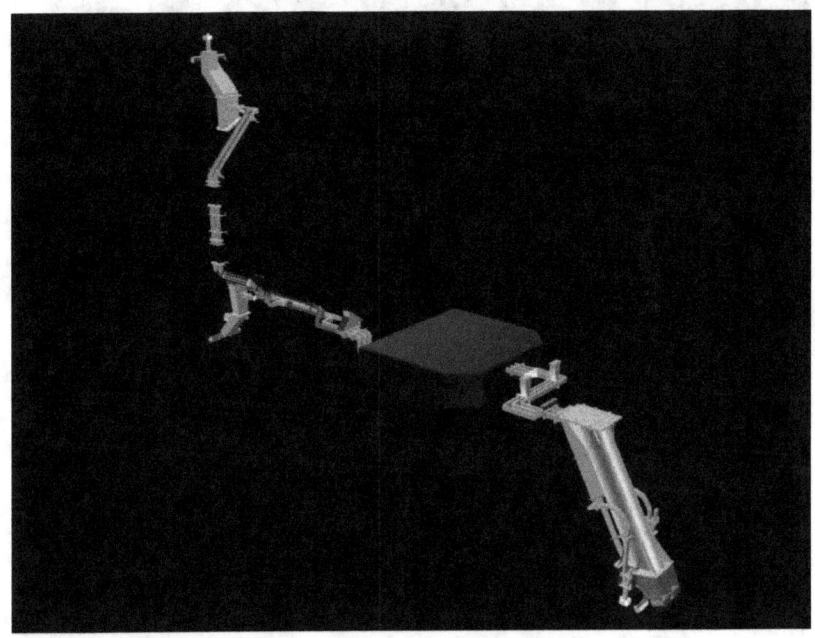

View 1 of Gravity/Radio Science. Image Credit: NASA.

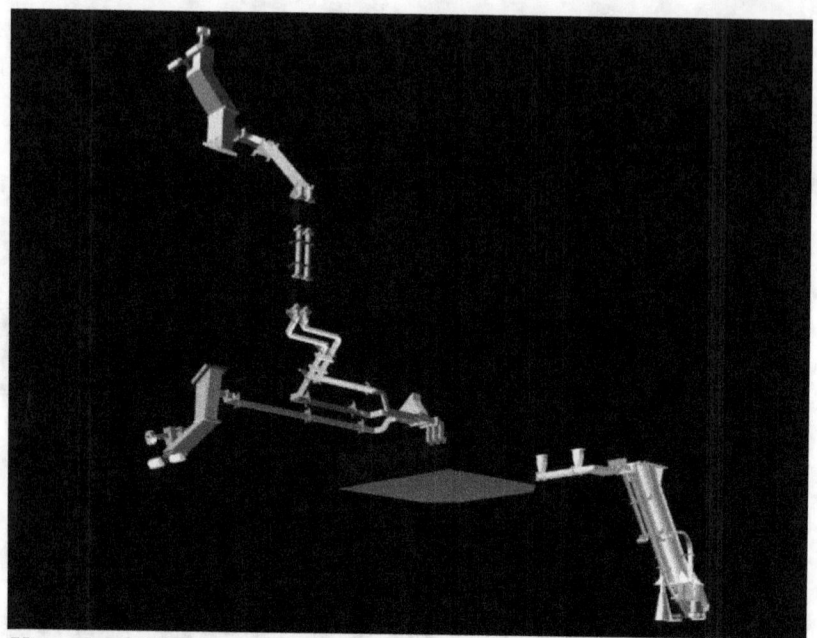
View 2 of Gravity/Radio Science. Image Credit: NASA.

REASON
Radar for Europa Assessment and Sounding: Ocean to Near-surface

Objective

Probe Europa's icy shell for the moon's suspected ocean and study the ice's structure and thickness. Also, study the moon's surface elevations, composition, and roughness, and search the moon's atmosphere for plumes.

Function

The pieces of the REASON experiment are mounted along the solar panel arms of the *Europa Clipper* spacecraft. REASON transmits radio waves that bounce off features within the underlying ice. Some radio waves return to the spacecraft, but at a fraction of their original energy. By measuring the time difference between transmission and return and knowing the speed radio waves travel through various materials, REASON learns how far the features are from the spacecraft.

By measuring the energy difference between transmitted signal and returning signal, combined with the measured distance, REASON can see differences in material properties. Scientists on Earth then use software to combine the radio waves to produce detailed images of the ice shell.

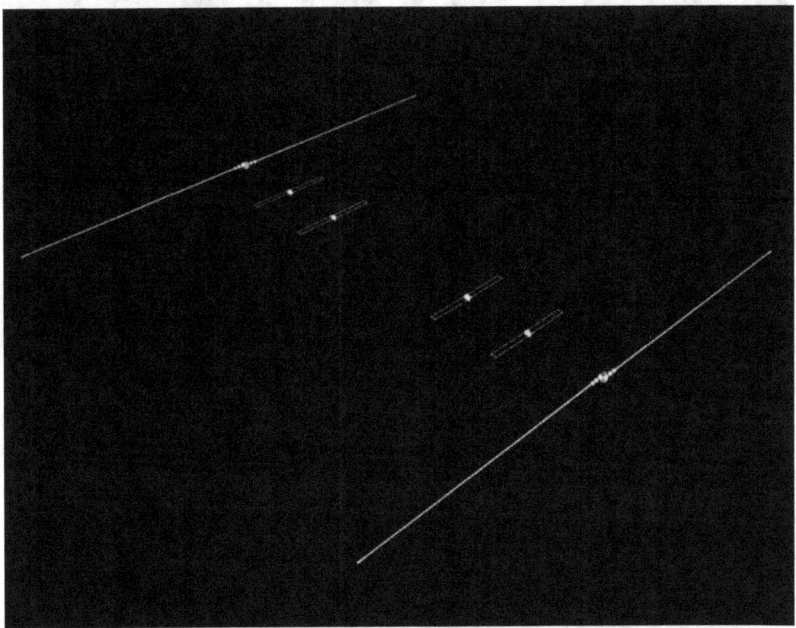

View 1 of Radar for Europa Assessment and Sounding: Ocean to Near-surface (REASON). Image Credit: NASA.

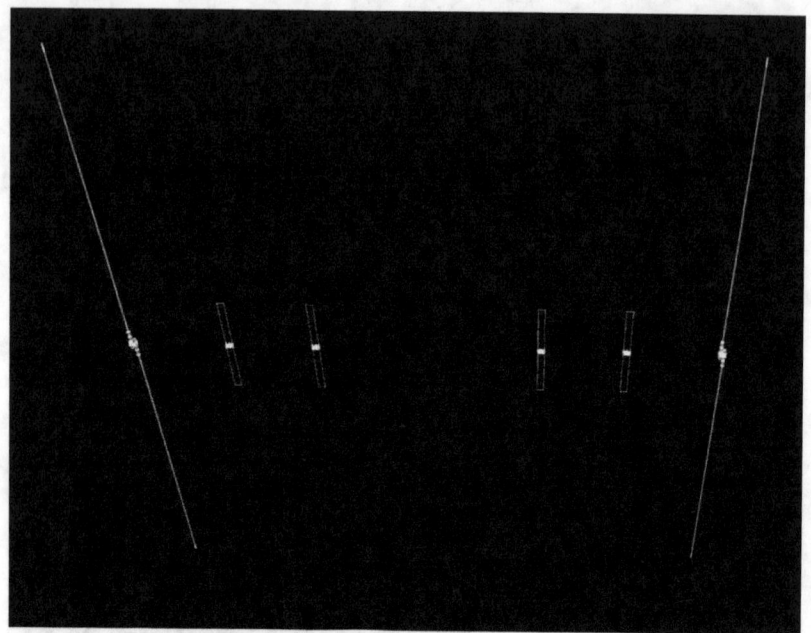

View 2 of Radar for Europa Assessment and Sounding: Ocean to Near-surface (REASON). Image Credit: NASA.

MASPEX
MAss Spectrometer for Planetary Exploration/Europa

Objective

Analyze the composition of Europa's thin atmosphere and any plumes of material emanating from the surface.

Function

MASPEX generates high-energy (fast moving) electrons to strip electrons from incoming gas molecules. That makes the gas molecules into positively-charged ions. The instrument accelerates the ions to a uniform amount of energy. The ionized gases are pulled into the "drift tube," which gives MASPEX its baguette-like length. The lighter the ion, the faster it can move through the drift tube.

MASPEX bounces the ions back and forth several times in the drift tube before the instrument detects them. The total distance they travel increases their difference in arrival time, magnifying their mass difference. It's like two siblings in a footrace. Racing across their backyard they might finish at almost the same time. But if they run around the block, their difference in speed is easier to observe.

By timing an ion that transit through the instrument, MASPEX determines the ions' mass. The mass reveals each molecule's identity which helps determine whether Europa is habitable.

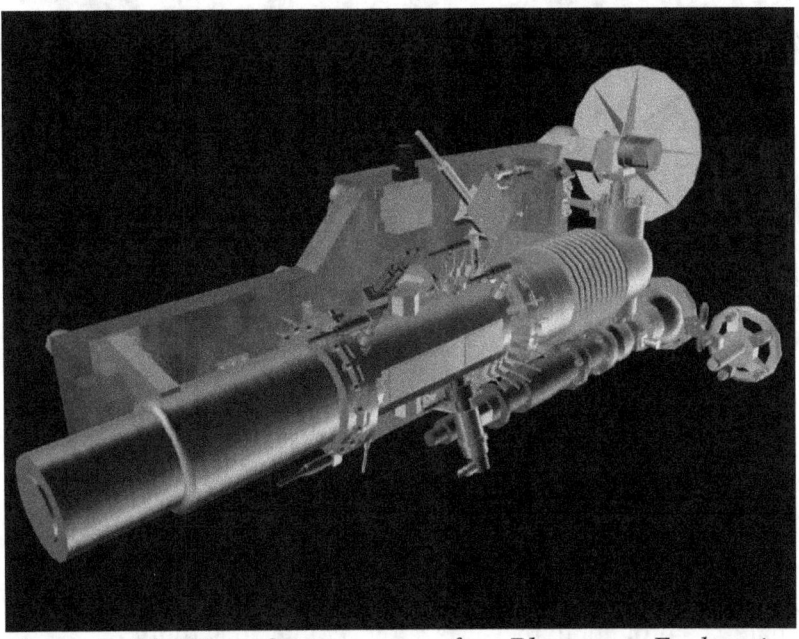

View 1 of Mass Spectrometer for Planetary Exploration (MASPEX). Image Credit: NASA.

View 2 of Mass Spectrometer for Planetary Exploration (MASPEX). Image Credit: NASA.

SUDA
SUrface Dust Analyzer

Objective

Analyze dust particles ejected from Europa's surface.

Function

Studying the dust particles that pass through SUDA helps identify the composition of Europa's surface and any subsurface materials brought to the surface by geological activity.

When dust enters SUDA, it passes through a series of metal mesh grids that measure the dust's speed and trajectory, which identify the dust's area of origin on Europa's surface.

The dust then strikes a metal target plate that shatters dust grains into individual modules and ionizes (electrically charges) some of them. Once ionized, the molecules must obey SUDA's electrical field, which funnels them to a detector. An ion's mass-to-charge ratio determines how long it takes to reach the detector. The timing reveals the molecule's mass and composition. This allows scientists to identify the material and also determine whether organic molecules are abiotic or biomolecules.

View 1 of Surface Dust Analyzer (SUDA). Image Credit: NASA.

View 2 of Surface Dust Analyzer (SUDA). Image Credit: NASA.

Integrating Data from Multiple Instruments

The instruments on *Europa Clipper* are designed to work together, providing complementary data that offers a comprehensive view of Europa's environment. By integrating data from multiple instruments, scientists can develop a detailed understanding of Europa's habitability. This approach allows for cross-verification of findings and helps build a robust scientific picture of Europa's potential to support life.

Expected Outcomes and Impact on Planetary Science

The experiments conducted by *Europa Clipper* are expected to yield significant scientific discoveries, including:

- Detailed maps of Europa's ice shell and subsurface ocean.

- Insights into the chemical composition of the surface and atmosphere.

- Identification of active geological processes and regions.

- Evidence of potential biosignatures or prebiotic chemistry.

These outcomes will have a profound impact on planetary science, advancing our understanding of icy moons and the potential for life beyond Earth. The data collected by *Europa Clipper* will not only answer key questions about Europa but also guide future missions, including potential landers or subsurface probes.

History of Jupiter System Exploration

Exploration of the Jupiter system has been a central focus of planetary science for decades. The journey to understand Jupiter and its moons has involved numerous missions, each building on the knowledge gained by its predecessors. In this chapter, we will trace the historical context of Jupiter system exploration, highlighting key missions and their contributions, and set the stage for the *Europa Clipper* mission within this rich legacy of discovery.

Early Missions: *Pioneer* and *Voyager*

***Pioneer* 10 and 11 (1973-1974):**

- *Pioneer* 10 became the first spacecraft to travel through the asteroid belt and make a close encounter with Jupiter in 1973. It provided the first direct observations of the planet and its moons, measuring Jupiter's radiation belts and magnetic field.

- *Pioneer* 11 followed in 1974, offering additional data and refining our understanding of Jupiter's environment. These missions laid the groundwork for subsequent, more detailed explorations.

***Voyager* 1 and 2 (1979):**

- The *Voyager* missions marked a significant leap forward in our understanding of the Jupiter system. In 1979, *Voyager* 1 and 2 conducted detailed flybys of Jupiter and its major moons—Io, Europa, Ganymede, and Callisto.

- They provided the first high-resolution images of these moons, revealing their diverse surfaces and geological features. Notably, *Voyager* 1 captured images of active volcanism on Io and intricate surface features on Europa, sparking interest in their geological and potentially astrobiological significance.

In-Depth Exploration: *Galileo* and *Cassini*

Galileo Orbiter (1995-2003):

- Launched in 1989, *Galileo* became the first spacecraft to orbit Jupiter, arriving in 1995. It conducted extensive studies of the planet and its moons over eight years.

- *Galileo* provided crucial data on Europa, including evidence of a subsurface ocean beneath its icy crust. It also discovered signs of potential liquid water in contact with the moon's surface materials, fueling speculation about its habitability.

- The mission also studied Io's volcanism, Ganymede's magnetic field, and Callisto's heavily cratered surface, significantly advancing our knowledge of the Jupiter system.

Cassini-Huygens (2000):

- While primarily a mission to Saturn, *Cassini* made a flyby of Jupiter in 2000, en route to its final destination. This flyby allowed for joint observations with *Galileo*, enhancing data collection on Jupiter's atmosphere and magnetosphere.

- *Cassini*'s observations complemented *Galileo*'s findings and provided a broader context for understanding Jupiter's influence on its moons and surrounding space environment.

Recent Missions: *Juno* and Future Endeavors

Juno (2016-Present):

- Launched in 2011 and arriving at Jupiter in 2016, *Juno* is currently (as of this writing) providing detailed studies of Jupiter's atmosphere, magnetic field, and interior structure.

- *Juno*'s close orbits and advanced instrumentation have revealed new insights into Jupiter's weather systems, polar cyclones, and deep atmospheric dynamics. It also offers data on the planet's gravitational and magnetic fields, contributing to our understanding of its internal composition.

- While *Juno* focuses primarily on Jupiter itself, its findings are essential for understanding the broader context in which Europa and the other moons exist.

Europa Clipper (2024):

- Launched October 14, 2024, *Europa Clipper* is the latest mission specifically designed to study one of Jupiter's moons in detail. Its focus on Europa builds directly on the discoveries made by *Pioneer*, *Voyager*, *Galileo*, and *Juno*.

- *Europa Clipper* aims to confirm the presence and characteristics of the subsurface ocean suggested by *Galileo*, using advanced instruments to probe the ice shell, ocean, and surface composition.

Significance of *Europa Clipper* in Historical Context

Europa Clipper represents the culmination of decades of exploration and scientific inquiry. Each preceding mission has contributed vital pieces of the puzzle, enhancing our understanding of the Jupiter system and refining the questions that *Europa Clipper* aims to answer. The mission's advanced technology and focused scientific goals position it to make groundbreaking discoveries about Europa's potential habitability.

Building on *Voyager* and *Galileo*:

- *Voyager*'s initial images of Europa's surface hinted at a dynamic world, while *Galileo*'s data provided strong evidence for a subsurface ocean. *Europa Clipper* builds on these findings, using sophisticated instruments to explore these features in unprecedented detail.

- The mission's ice-penetrating radar, magnetometer, and other instruments are designed to answer the questions raised by previous missions: How thick is the ice shell? What is the composition of the ocean? Is there evidence of current geological activity?

Complementing *Juno*:

- While *Juno* provides critical data on Jupiter's atmosphere and magnetic environment, *Europa Clipper* will focus on Europa itself. The combined data from both missions will offer a comprehensive view of how Jupiter influences its moons and how these moons, in turn, respond to and interact with their environment.

Setting the Stage for Future Exploration:

- The discoveries made by *Europa Clipper* will inform and inspire future missions, including potential landers or probes designed to explore Europa's surface and subsurface directly. By characterizing Europa's environment and identifying regions of interest, *Europa Clipper* paves the way for the next generation of exploration.

In the grand narrative of Jupiter system exploration, *Europa Clipper* stands as a pivotal chapter, poised to transform our understanding of this enigmatic moon and its potential for life. As we look to the future, the mission promises to deepen our knowledge of the solar system and our place within it, continuing humanity's quest to explore and understand the cosmos.

The Future of Europa Exploration

With the October 14, 2024-launch of the *Europa Clipper* mission, and eventual arrival at Europa, the scientific community is already envisioning the next steps in the exploration of this enigmatic moon. The data and insights gathered by *Europa Clipper* will set the stage for a new era of exploration, marked by increasingly sophisticated missions designed to answer the most profound questions about Europa's potential to support life. This chapter explores the future of Europa exploration, including potential follow-up missions, technological advancements, and the broader implications for planetary science and astrobiology.

Follow-Up Missions

1. **Europa Lander:**

- **Concept and Objectives:**

  - A proposed Europa Lander mission aims to complement the *Europa Clipper* by conducting in situ analysis of Europa's surface. The lander would be equipped with instruments to sample and analyze the ice, searching for biosignatures and characterizing the chemical composition directly.

  - Key objectives include drilling into the ice to access subsurface material, analyzing its physical and chemical properties, and identifying organic compounds or other indicators of biological activity.

- **Challenges and Technological Innovations:**

  - Landing on Europa poses significant challenges due to its icy and potentially unstable surface. Advanced landing technologies, including autonomous hazard avoidance and precision landing capabilities, are critical for the success of such a mission.

- The lander would also require robust systems to withstand Europa's harsh radiation environment and extreme temperatures.

2. **Subsurface Probes and Cryobots:**

- **Concept and Objectives:**

  - To explore the subsurface ocean directly, future missions may involve cryobots—robotic probes designed to melt through the ice shell and access the liquid water beneath. These probes could carry instruments to measure the ocean's properties and search for microbial life.

  - Objectives include assessing the ocean's depth, salinity, temperature, and potential hydrothermal activity, which could provide energy sources for life.

- **Challenges and Technological Innovations:**

  - Developing cryobots capable of penetrating the thick ice shell, estimated to be several kilometers deep, is a formidable engineering challenge. The probes must be equipped with autonomous navigation systems and efficient power sources to maintain operations during the descent.

  - Communication with the surface or an orbiting spacecraft is another challenge, necessitating advanced data relay systems or tethered connections.

Technological Advancements

1. **Improved Instrumentation:**

- Future missions will benefit from advancements in miniaturization and sensitivity of scientific instruments. Enhanced spectrometers, mass analyzers, and imaging systems will enable more detailed and precise measurements of Europa's environment.

- Innovations in sample collection and analysis technologies will allow for more comprehensive studies of surface and subsurface materials, increasing the likelihood of detecting biosignatures.

2. **Autonomous Systems and AI:**

- Autonomous systems and artificial intelligence (AI) will play a crucial role in future Europa missions. AI-driven data analysis, decision-making, and navigation will enhance mission efficiency and responsiveness to unexpected conditions.

- Autonomous rovers and probes will be able to conduct extended scientific investigations with minimal intervention from Earth, increasing the scientific return of each mission.

3. **Radiation Protection:**

- Advancements in radiation-hardened electronics and shielding materials will be essential for protecting spacecraft systems and instruments from Jupiter's intense radiation environment. Improved radiation protection will extend mission lifespans and enable longer-duration studies.

Broader Implications for Planetary Science and Astrobiology

1. **Understanding Habitability:**

- The exploration of Europa will significantly advance our understanding of habitability in the solar system. By studying Europa's ice shell, subsurface ocean, and potential hydrothermal activity, scientists can identify the key factors that make an environment suitable for life.

- Insights gained from Europa will inform the search for habitable environments on other icy moons, such as Enceladus and Titan, and guide the development of criteria for assessing exoplanet habitability.

2. **Advancing Astrobiology:**

- Discovering signs of life or prebiotic chemistry on Europa would be one of the most profound scientific achievements of our time. Such findings would have far-reaching implications for our understanding of life's origins, distribution, and potential diversity in the universe.

- Europa exploration will drive advancements in astrobiology, including the development of new life-detection technologies and the refinement of models for life's potential evolution in extreme environments.

3. **Inspiring Future Exploration:**

- The *Europa Clipper* mission and its successors will inspire a new generation of scientists, engineers, and explorers. The challenges and triumphs of these missions will captivate public imagination and foster a sense of wonder about the cosmos.

- Educational and outreach initiatives associated with Europa exploration will promote STEM (Science, Technology, Engineering, and Mathematics) education and encourage young people to pursue careers in space science and exploration.

The future of Europa exploration is filled with promise and excitement. As the *Europa Clipper* mission prepares to embark on its journey, the scientific community is already looking ahead to the next steps in unraveling the mysteries of this fascinating moon. From landers and cryobots to advanced instrumentation and autonomous systems, the technologies and missions on the horizon will build on the foundation laid by *Europa Clipper*, driving new discoveries and expanding our understanding of the solar system.

The exploration of Europa is more than a quest for scientific knowledge; it is a journey that embodies humanity's enduring spirit of curiosity and exploration. As we stand on the brink of this new era, we are reminded that the search for life beyond Earth is not just about finding other worlds—it is about discovering our place in the universe and understanding the fundamental nature of life itself. The future of Europa exploration holds the potential to transform our view of the cosmos and inspire generations to come.

Conclusion

The journey to explore Jupiter's moon Europa represents one of the most ambitious and scientifically promising endeavors in the history of space exploration. The *Europa Clipper* mission stands at the forefront of this quest, equipped with advanced technology and driven by profound scientific questions about the potential for life beyond Earth. As we conclude this exploration of the *Europa Clipper* mission and its context, it is essential to reflect on the broader implications of this journey and what it means for the future of space exploration and humanity's understanding of the cosmos.

The Significance of *Europa Clipper*

Europa Clipper is poised to revolutionize our understanding of one of the most intriguing worlds in our solar system. Its suite of sophisticated instruments is designed to probe beneath Europa's icy surface, map its subsurface ocean, and analyze the moon's geology and chemistry. The mission's primary objectives—assessing Europa's habitability, understanding its ice shell, and searching for signs of recent geological activity—are crucial for answering fundamental questions about the potential for life elsewhere in the solar system.

The data collected by *Europa Clipper* will not only advance our knowledge of Europa but also provide critical insights into the broader processes that shape icy moons and other celestial bodies. By studying Europa's environment, scientists can develop better models for understanding the habitability of other icy worlds, both within our solar system and in distant planetary systems.

Building on a Legacy of Exploration

The *Europa Clipper* mission builds on a rich legacy of exploration that includes pioneering missions such as *Pioneer*,

Voyager, *Galileo*, and *Juno*. Each of these missions has contributed essential knowledge about Jupiter and its moons, setting the stage for *Europa Clipper*'s focused study of Europa. This continuity of exploration underscores the collaborative and cumulative nature of space science, where each mission builds on the discoveries of its predecessors.

Europa Clipper also represents a bridge to future missions. The detailed reconnaissance provided by Clipper will inform the design and objectives of subsequent missions, such as potential landers and subsurface probes, which aim to delve even deeper into Europa's secrets. These future missions will continue the scientific journey, pushing the boundaries of what we know and can achieve.

Implications for Astrobiology and Planetary Science

The exploration of Europa is at the forefront of astrobiology—a field dedicated to understanding the potential for life beyond Earth. Discovering signs of life or prebiotic chemistry on Europa would have profound implications for our understanding of life's origins, evolution, and distribution in the universe. It would challenge and expand our knowledge of biology, chemistry, and the environmental conditions that can support life.

In a broader sense, studying Europa and other icy moons helps scientists understand the processes that govern the formation and evolution of planetary systems. These insights are crucial for interpreting observations of exoplanets and their potential habitability. *Europa Clipper*'s findings will thus contribute to our understanding of the cosmos and our place within it.

Inspiring Future Generations

Space exploration has always had the power to inspire and unite people across the globe. The *Europa Clipper* mission is no exception. Its journey to one of the most distant and mysterious moons in our solar system captures the imagination and fuels a sense of wonder and curiosity. This inspiration extends beyond the scientific community to the public, fostering a greater appreciation for science, technology, engineering, and mathematics (STEM) fields.

Educational and outreach initiatives associated with *Europa Clipper* will engage students, educators, and the general public, encouraging a new generation of scientists, engineers, and explorers. The mission's discoveries and the stories of the people who make them possible will inspire future leaders in space exploration and other fields.

A New Era of Exploration

As we stand on the brink of a new era in the exploration of Europa, it is clear that the *Europa Clipper* mission is more than just a scientific endeavor. It is a testament to human ingenuity, curiosity, and the relentless pursuit of knowledge. The mission embodies the spirit of exploration that has driven humanity to seek out new frontiers, push the boundaries of what is possible, and strive for a deeper understanding of the universe.

In the coming years, as *Europa Clipper* embarks on its journey and begins to transmit its findings back to Earth, we will undoubtedly be faced with new questions, challenges, and opportunities. The mission will likely open doors to further exploration and discovery, continuing a legacy of scientific inquiry that extends far into the future.

The exploration of Europa is a journey that holds the promise of transforming our understanding of the solar system and our place within it. As we conclude this book, we are reminded that the quest for knowledge is an ongoing adventure—one that takes us to the farthest reaches of space and into the deepest mysteries of existence. The *Europa Clipper* mission is a significant milestone on this journey, and its legacy will inspire and guide future generations as they continue to explore the wonders of the cosmos.

Appendices

Glossary of Scientific Terms

Astrobiology
The study of the origin, evolution, distribution, and future of life in the universe. Astrobiology seeks to understand the potential for life on other planets and moons, including Europa.

Cryobot
A robotic probe designed to melt through ice to reach and explore subsurface environments, such as the ocean believed to exist beneath Europa's ice shell.

Exosphere
The outermost layer of an atmosphere, where particles are so sparse that they can travel significant distances without colliding with one another. Europa has a tenuous exosphere composed mainly of oxygen.

Habitability
The potential of an environment to support life. For Europa, this includes factors such as the presence of liquid water, energy sources, and the right chemical conditions.

Hydrothermal Activity
Geological activity involving the interaction of water and heat, often resulting in the formation of hydrothermal vents. Such activity on Europa could provide energy sources for potential life.

Magnetosphere
The region around a planet dominated by its magnetic field. Jupiter's strong magnetosphere interacts with its moons, including Europa, affecting their environments.

Radiation Belts
Zones of charged particles trapped by a planet's magnetic field. Jupiter's intense radiation belts pose significant challenges for spacecraft and potential future human exploration.

Spectrometer
An instrument that measures the properties of light over a specific portion of the electromagnetic spectrum, often used to identify the composition of materials.

Detailed Mission Timeline

The *Europa Clipper* mission, from its initial conception to its anticipated execution, represents a monumental effort in space exploration. This timeline highlights key events in the mission's development and planned milestones, showcasing the extensive planning, technological innovation, and scientific collaboration that have brought the mission to fruition.

Idea, Early Concepts, and Mission Planning

2008: Mission Idea

Event: The idea of a dedicated mission to Europa emerged from the scientific community, and interest grew over time.

2013: Initial Mission Concept

Event: NASA begins considering mission concepts for exploring Europa, focusing on the potential for habitability and the search for life.

Significance: This marks the start of serious discussions and feasibility studies for a dedicated mission to Europa.

2014: Science Definition Team Report

Event: The Europa Science Definition Team publishes a report outlining key scientific objectives and mission goals.

Significance: This report forms the basis for the mission's scientific framework and priorities.

2015: Mission Approval and Funding

Event: NASA formally approves the *Europa Clipper* mission and secures initial funding from Congress.

Significance: The mission receives official backing, allowing detailed planning and development to proceed.

Design and Development Phase

2016: Preliminary Design Review (PDR)

Event: The *Europa Clipper* mission passes its Preliminary Design Review, a critical milestone in mission development.

Significance: Approval at this stage ensures that the mission design meets required standards and is feasible within the set budget and timeline.

2017: Selection of Scientific Instruments

Event: NASA selects nine scientific instruments for the *Europa Clipper* spacecraft, including radar, spectrometers, and imagers.

Significance: These instruments are chosen to address key scientific questions about Europa's habitability and geology.

2019: Final Design and Fabrication

Event: The mission enters the final design and fabrication phase, with spacecraft components being built and tested.

Significance: This phase is crucial for assembling and integrating the spacecraft's systems and instruments.

Testing and Integration

2020: Assembly, Test, and Launch Operations (ATLO)

Event: The spacecraft enters the Assembly, Test, and

Launch Operations phase, where components are integrated and thoroughly tested.

Significance: Ensuring that all systems function correctly in simulated space conditions is vital for mission success.

2021: Environmental Testing

Event: The *Europa Clipper* undergoes rigorous environmental testing, including thermal, vacuum, and vibration tests.

Significance: These tests simulate the harsh conditions of space to ensure the spacecraft can withstand the journey to Jupiter and operations in its environment.

2022: Final Integration

Event: Final integration of the spacecraft and its instruments is completed.

Significance: The fully assembled spacecraft is prepared for pre-launch activities, including system checks and final calibrations.

Pre-Launch Preparations

2023: Shipment to Launch Site

Event: The spacecraft is transported to the Kennedy Space Center in Florida, the designated launch site.

Significance: This marks the final step in preparing the spacecraft for its journey to Europa.

2024: Launch Readiness Review

Event: NASA conducts the Launch Readiness Review to

ensure all systems are go for launch.

Significance: A successful review confirms that the spacecraft and launch vehicle are ready for liftoff.

Launch and Cruise Phase

October 14, 2024, at 1:26 p.m. EDT: Launched

Event: Launched aboard a SpaceX Falcon Heavy rocket from Launch Pad 39A at Kennedy Space Center, the spacecraft begins its journey to Jupiter, marking a major milestone in NASA's exploration of the outer solar system.

Significance: The successful launch is the culmination of years of preparation, and it sets the stage for one of the most important scientific missions of the coming decade.

2024-2025: Initial Spacecraft Checkout and Instrument Activation

Event: Following the launch, *Europa Clipper* undergoes a series of post-launch tests and checks, ensuring all systems and scientific instruments are fully operational as it travels through interplanetary space.

Significance: This phase ensures that the spacecraft is in full working order for its long journey to Jupiter.

March 1, 2025: Mars Flyby

Event: The spacecraft completed its first gravity-assist maneuver.

Significance: The first of two maneuvers essential for efficiently reaching Jupiter and setting up the desired orbital path around the planet.

December 3, 2026: Earth Flyby

Event: The spacecraft will complete its second gravity-assist maneuver.

Significance: The second of two maneuvers essential for efficiently reaching Jupiter and setting up the desired orbital path around the planet.

2025-2030: Cruise to Jupiter

Event: Following the two gravity-assist maneuvers, the spacecraft will follow a high-energy arc outward, eventually arriving at Jupiter in April of 2030.

Significance: These maneuvers are essential for efficiently reaching Jupiter and setting up the desired orbital path around the planet.

Arrival and Operational Phase

April 2030: Arrival at Jupiter

Event: The *Europa Clipper* is expected to arrive at Jupiter and begin its scientific mission.

Significance: Upon arrival, the spacecraft will enter an elliptical orbit around Jupiter, positioning it for close flybys of Europa.

2030-2036: Europa Flybys and Data Collection

Event: The *Europa Clipper* will conduct approximately 45 close flybys of Europa, collecting high-resolution data on the moon's surface and subsurface.

Significance: These flybys will enable detailed mapping and analysis of Europa's ice shell, ocean, and potential plumes, providing critical data to assess its habitability.

2036: End of Primary Mission

Event: The primary mission phase is expected to conclude, depending on the spacecraft's health and the availability of extended mission funding.

Significance: The end of the primary mission will mark the culmination of years of planning and exploration, with the potential for extended missions based on the spacecraft's condition and scientific priorities.

Conclusion

The *Europa Clipper* mission timeline reflects the extensive effort and collaboration required to undertake such a complex and ambitious project. From initial concepts to the anticipated arrival at Jupiter, each milestone represents a significant achievement in space exploration. As the mission progresses, it promises to yield groundbreaking discoveries about Europa and the potential for life beyond Earth, inspiring future generations and advancing our understanding of the solar system.

References and Further Reading

This section provides a comprehensive list of references and suggested readings for those interested in exploring more about the *Europa Clipper* mission, the scientific objectives, and the broader context of space exploration. These resources include scientific papers, mission documentation, books, and online articles that offer in-depth information and insights.

Key Scientific Papers and Reports

Pappalardo, R. T., et al. (2013). "Science Potential from a Europa Lander." *Astrobiology*, 13(8), 740-773.

> Summary: This paper discusses the scientific potential and goals of a future Europa lander mission, complementing the objectives of the *Europa Clipper* mission.
>
> Link: https://doi.org/10.1089/ast.2013.1003

Phillips, C. B., & Pappalardo, R. T. (2014). "*Europa Clipper* Mission: Exploring Jupiter's Ocean Moon." *Eos, Transactions American Geophysical Union*, 95(20), 165-167.

> Summary: An overview of the *Europa Clipper* mission, detailing its scientific goals and expected contributions to our understanding of Europa.
>
> Link: https://agupubs.onlinelibrary.wiley.com/doi/full/10.1002/2014EO200002

Hand, K. P., et al. (2017). "Europa Lander Study 2016 Report: Science of Europa Exploration." NASA Technical Reports Server.

> Summary: A comprehensive report on the scientific objectives and potential payloads for a Europa lander

mission, highlighting synergies with the *Europa Clipper* mission.

Books

Greenberg, R. (2005). *Unmasking Europa: The Search for Life on Jupiter's Ocean Moon*. Springer.

Summary: This book provides an in-depth look at Europa, exploring its potential for harboring life and the scientific missions aimed at studying this intriguing moon.

Schmidt, B. E., & Blankenship, D. D. (Eds.). (2018). *Europa: The Ocean Moon*. Cambridge University Press.

Summary: A collection of essays and studies by leading scientists, offering detailed insights into Europa's geology, ice shell, ocean, and the technological challenges of exploring this moon.

Online Resources

NASA's *Europa Clipper* Mission Page

Description: The official NASA page for the *Europa Clipper* mission, providing up-to-date information on mission status, scientific objectives, and detailed descriptions of the spacecraft and its instruments.

Link: https://europa.nasa.gov

Jet Propulsion Laboratory (JPL) - *Europa Clipper*

Description: JPL's dedicated *Europa Clipper* page, featuring news, mission updates, and educational resources about the mission and Europa itself.

Link: https://www.jpl.nasa.gov/missions/europa-clipper

The Planetary Society - *Europa Clipper* Mission

Description: An overview of the *Europa Clipper* mission, including its scientific objectives, mission timeline, and significance within the broader context of planetary exploration.

Link: https://www.planetary.org/space-missions/europa-clipper

Further Reading

Pappalardo, R. T., & Barr, A. C. (2004). "Europa's Icy Shell: Past, Present, and Future." Eos, Transactions American Geophysical Union, 85(18), 177-181.

Summary: This paper provides a comprehensive overview of Europa's icy shell, discussing its formation, current state, and the scientific questions that future missions like *Europa Clipper* aim to answer.

Link: https://agupubs.onlinelibrary.wiley.com/doi/abs/10.1029/2004EO180001

These references and further reading materials provide a solid foundation for anyone interested in the *Europa Clipper* mission and the broader context of exploring Jupiter's moons. Whether you are a student, researcher, or space enthusiast, these resources offer valuable insights and in-depth information about one of the most exciting missions in contemporary space exploration.

References

The following references provide a comprehensive foundation for the material discussed in this book. They include scientific papers, mission reports, and authoritative sources on space exploration, specifically focusing on Europa, the *Europa Clipper* mission, and the broader context of Jupiter system exploration.

NASA (2017). *Europa Clipper Mission Overview.* Retrieved from NASA. https://europa.nasa.gov/

> An overview of the *Europa Clipper* mission, detailing its objectives, scientific instruments, and mission timeline.

Pappalardo, R. T., et al. (2013). "Science Potential from a Europa Lander." *Astrobiology*, 13(8), 740-773. https://doi.org/10.1089/ast.2013.1003

> Discusses the scientific potential and goals of a future Europa lander mission, complementing the objectives of the *Europa Clipper* mission.

Phillips, C. B., & Pappalardo, R. T. (2014). "*Europa Clipper* Mission: Exploring Jupiter's Ocean Moon." *Eos, Transactions American Geophysical Union*, 95(20), 165-167. https://agupubs.onlinelibrary.wiley.com/doi/full/10.1002/2014EO200002

> An overview of the *Europa Clipper* mission, detailing its scientific goals and expected contributions to our understanding of Europa.

Greenberg, R. (2005). *Unmasking Europa: The Search for Life on Jupiter's Ocean Moon.* Springer.

> Provides an in-depth look at Europa, exploring its potential for harboring life and the scientific missions

aimed at studying this intriguing moon.

Schmidt, B. E., & Blankenship, D. D. (Eds.). (2018). *Europa: The Ocean Moon.* Cambridge University Press.

> A collection of essays and studies by leading scientists, offering detailed insights into Europa's geology, ice shell, ocean, and the technological challenges of exploring this moon.

NASA Jet Propulsion Laboratory (JPL). *Europa Clipper Mission Page.* Retrieved from JPL. https://www.jpl.nasa.gov/missions/europa-clipper

> JPL's dedicated Europa Clipper page, featuring news, mission updates, and educational resources about the mission and Europa itself.

The Planetary Society. *Europa Clipper Mission Overview.* Retrieved from The Planetary Society. https://www.planetary.org/space-missions/europa-clipper

> An overview of the *Europa Clipper* mission, including its scientific objectives, mission timeline, and significance within the broader context of planetary exploration.

Pappalardo, R. T., & Barr, A. C. (2004). "Europa's Icy Shell: Past, Present, and Future." *Eos, Transactions American Geophysical Union*, 85(18), 177-181. http://dx.doi.org/10.1016/j.icarus.2005.08.002

> Provides a comprehensive overview of Europa's icy shell, discussing its formation, current state, and the scientific questions that future missions like *Europa Clipper* aim to answer.

About the Author

James M. Thomas was born and raised in Tampa, Florida, and lives there with his family. Thomas has many and varied interests and tries to dabble in as many of them as time permits.

www.ingramcontent.com/pod-product-compliance
Lightning Source LLC
Chambersburg PA
CBHW071840210526
45479CB00001B/219